Health 94

你真聪明

How Smart Are You

Gunter Pauli

[比] 冈特·鲍利 著

[哥伦] 凯瑟琳娜·巴赫 绘

何家振 译

上海远东出版社

特别感谢以下热心人士对童书工作的支持：

匡志强　宋小华　解　东　厉　云　李　婧　庞英元
李　阳　刘　丹　冯家宝　熊彩虹　罗淑怡　旷　婉
杨　荣　刘学振　何圣霖　廖清州　谭燕宁　王　征
李　杰　韦小宏　欧　亮　陈强林　陈　果　寿颖慧
罗　佳　傅　俊　白永喆　戴　虹

目录

Contents

一只章鱼正在抱怨有些人喜欢活吃他们。一只牡蛎坐在一旁，倾听章鱼冗长枯燥的抱怨。

“你以为只有你们章鱼会被活吃吗？”牡蛎问道，“我们也经常被人活吃呢。”

An octopus is complaining that some people like to eat him alive. An oyster sitting nearby listens to this litany of complaints from the octopus.

“So you think you are the only one eaten alive?” asks the oyster. “I am also devoured by people while I am still kicking.”

一只章鱼正在抱怨

An octopus is complaining

你真的有很多腿啊！

You do have an awful lot of legs!

“我不知道，为什么把我杀死，而且在我的肢体还在蠕动的时候咀嚼它，会给人们带来快乐。”章鱼抱怨道。

“你真的有很多腿啊！”

“I don’t know why it gives people joy to slaughter me and then start chewing on my limbs while I am still moving them,” complains the octopus.

“You do have an awful lot of legs!”

“我原谅你把我的胳膊说成‘腿’。我其实是有四对胳膊，就是一共八条胳膊。”

“你被认为是这附近最聪明的动物之一！”

“我并不认为我有那么聪明，但是如果有只螃蟹藏在箱子里，人们应该知道我会钻进那只箱子。如果我知道船上装满了螃蟹，我会跳上那条船——当然了，还会吃掉那些螃蟹。”

“You are forgiven for thinking that I have legs. I really have four pairs of arms. So that makes it eight arms in total.”

“And you are supposed to be one of the smartest animals around!”

“I don’t think I’m that smart, but if there is a crab in a box then people had better know that I will get inside that box. I will even jump on a boat when I know it’s full of crabs – and eat them, of course.”

我其实是有四对胳膊

I really have four pairs of arms

如果你被困在箱子里怎么办？

And what if you are trapped in a box?

“如果你被困在箱子里怎么办？”

“只要我能发现一个2厘米的洞，我就能钻出来!”

“不可能，你太大了！如果你试图钻出来，你的背部会被挤断的。”

“And what if you are trapped in a box?”

“If I can find a 2 centimetre hole, then I will get out!”

“Impossible, you are way too big! And if you ever try you will break your back.”

“我根本没有脊骨——这就是为什么我会如此柔软。”

“你不像我一样有很硬的外壳。那你是咬个洞钻出来吗？”

“虽然我的舌头上长满了牙齿，但我不是靠它出来的。只要能将我的胳膊一条一条挤过去，我就可以从任何空隙钻出来。”

“可是，你的头怎么办呢?”

“I do not have a backbone – that’s what makes me so flexible.”

“You do not have a shell like me either. So will you bite your way through it?”

“My tongue is full of teeth, but I do not need them for that. I can squeeze through any space provided I can get one arm through after the other.”

“And what about your head?”

我的舌头上长满了牙齿

My tongue is full of teeth

我的头没有问题

My head is no problem

“我的头没有问题，别忘了我没有脊椎、没有颅骨。我得确保我的三颗心脏折叠好，只要它们能通过就没问题了。”

“你一定是在开玩笑吧，没人有三颗心脏。”

“My head is no problem; remember I have no back and no skull. I must simply ensure that my three hearts can be folded so that they get through.”

“You’ve got to be joking, no one has three hearts.”

“你为什么不相信呢？世界之大，无奇不有，总会有些例外。我需要一个额外的心脏给我的八条胳膊供血。”

“一颗心脏专门为你的胳膊供血？那另外两颗心脏是干吗的？”

“我的双鳃。”

“对不起，我只是一只牡蛎——我不明白，为什么你的两个鳃需要两个心脏？”

“Why can’t you believe this? Nature is full of surprises, and there are always exceptions. I need an extra heart to pump blood through my eight arms.”

“An extra one just for your arms? And what are the other two hearts used for?”

“My gills.”

“Sorry, but I am just an oyster – why do you need two hearts for two gills?”

世界之大，无奇不有

Nature is full of surprises

我的鳃就是我的肺

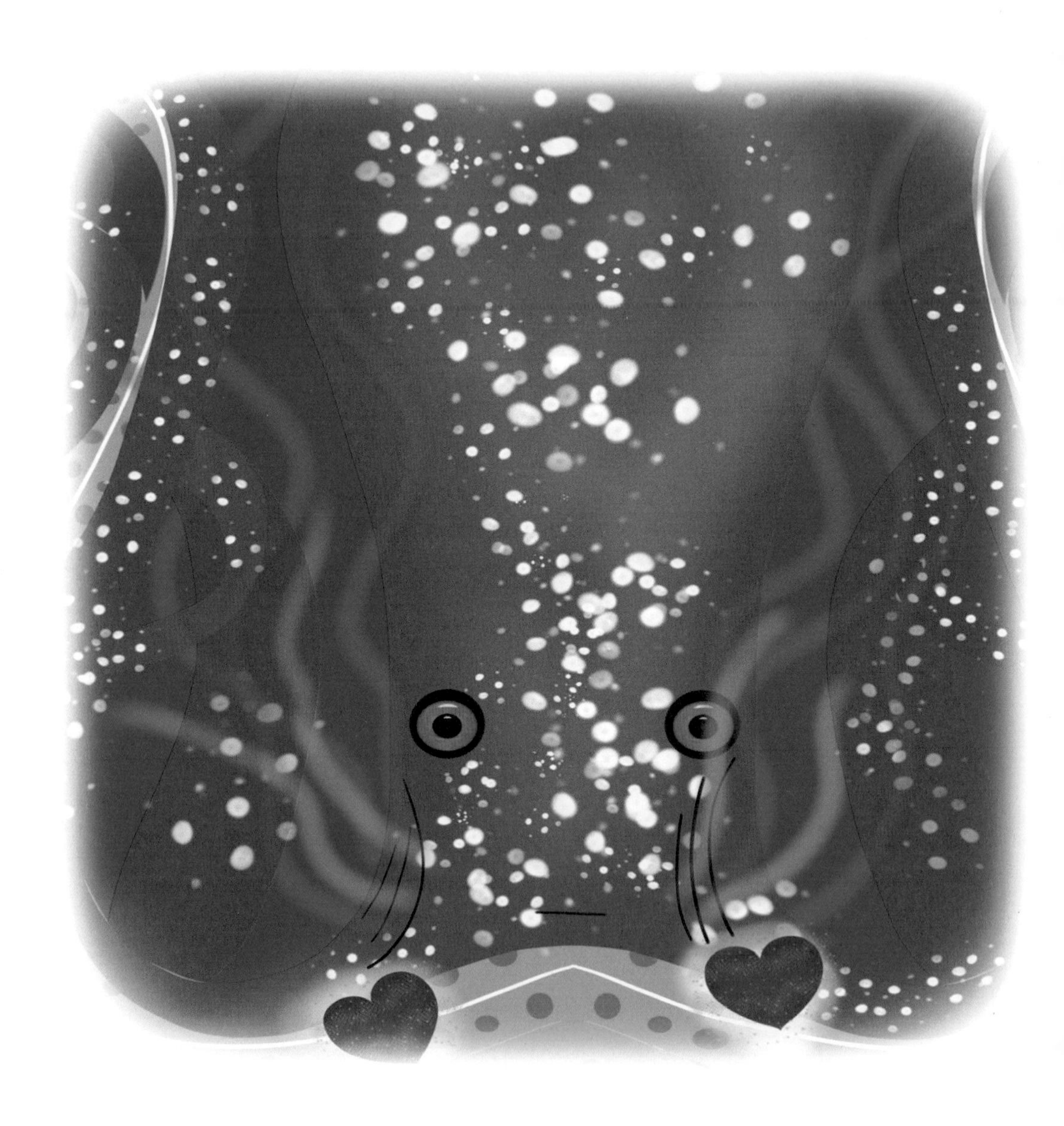

My gills are my lungs

“我的鳃就是我的肺。我从水中吸收氧气，再把二氧化碳排进水中。这需要很多能量，特别是当我在4 000米深的海底游动时，那里的水中几乎没有氧气。”

“在水中呼吸一点也不稀奇，我们都是在水里呼吸的。但是，如果你断掉一条胳膊，真的还会再长出来吗？”

“My gills are my lungs. I take oxygen out of the water, and pump carbon dioxide back into the water. This requires a lot of energy, especially when you are swimming 4,000 metres deep, where there is hardly any oxygen in the water.”

“Breathing in water is nothing special, we all do that. But is it true that if you rip off one of your arms it will grow again?”

“当然，而且在遇到危险时，我还会变颜色呢。如果需要，我们还能改变水的颜色，以便逃跑。”

“难怪人们那么喜欢活吃你呢！”

“人们最好明白，即使将我切碎，我仍然能钻回他们的喉咙。他们甚至会因此窒息！我已经下决心要反抗到底。”

……这仅仅是开始！……

“Sure, and when there is danger around, I will change colour. If we need to, we can change the colour of the water too, allowing us to escape.”

“You sure make these people run for the pleasure of eating you alive!”

“People had better realise that even if they cut me up, I will crawl back up their throat. They may even choke! I am ready to fight until the very bitter end.”

... AND IT HAS ONLY JUST BEGUN!...

……这仅仅是开始！……

… AND IT HAS ONLY JUST BEGUN! …

Did You Know?

你知道吗？

There are 289 different types of octopus, all living in salt water. The biggest one weighs up to 300 kg and the smallest only weighs one gram. All of them have toxic ink.

有289种不同的章鱼，它们全都居住在海水之中。最大的章鱼重300千克，而最小的章鱼只有1克重。它们都能喷出毒汁。

The teeth of the octopus are not on its jaw, since it has no skull. They are located on its tongue. An octopus can remain without food for 6 months.

因为章鱼没有颅骨，它的牙不是长在下巴上，而是长在舌头上。章鱼不吃食物也能生存6个月。

Octopuses' favourite food is crabs. They crush it to pieces in their beaks, which are made of keratin – the same material as human fingernails.

章鱼最喜欢的食物是螃蟹，它们用腭把螃蟹碾碎。章鱼的腭是角质的，成分类似于人类的指甲。

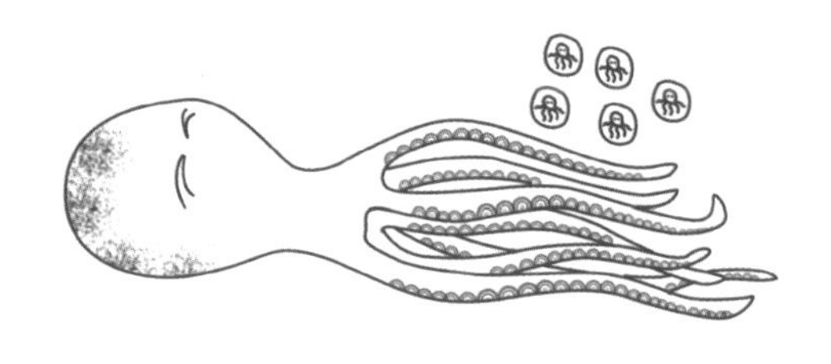

The eyes of an octopus focus like a camera, moving the lens in and out. They have no blind spot, and have polarised vision, meaning that they can clearly see jellyfish, which we can hardly see.

章鱼眼睛就像一个照相机镜头，能够伸缩调焦。章鱼眼睛没有盲点，有极好的视力，能够清楚地看到人类很难看到的水母。

Octopuses have excellent memory, and their brain has folds, indicating its complexity. The octopus can unscrew jars to get to the crabs inside.

章鱼有很强的记忆力，它们的大脑具有沟回，这意味着它们的大脑很复杂。章鱼能够拧开坛子的盖子，去抓里面的螃蟹。

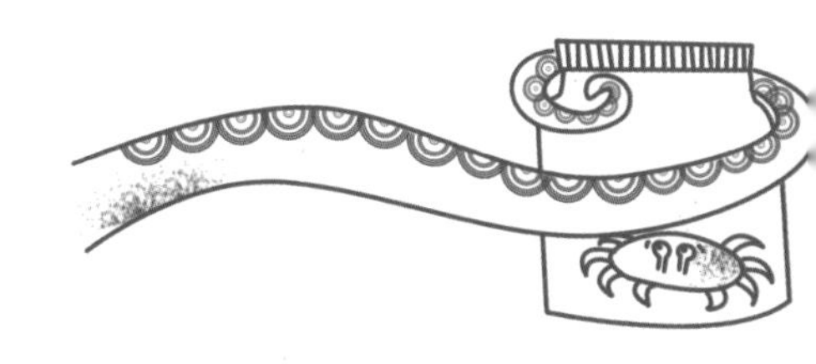

The octopus is a very efficient feeder. Multiply its weight by three and then you know how much it has eaten during its lifetime.

章鱼是非常高效的吃货。吃饱时其体重会增加三倍。由此你可以想象，章鱼的一生吃掉了多少东西。

The octopus has a short life span, with a maximum of 5 years. The octopus stops eating after breeding and starves to death, dying shortly after its eggs have hatched.

章鱼的生命周期很短，寿命最长的章鱼只能活 5 年。章鱼生育后就停止进食，直到饿死，在它的蛋孵出后不久就死掉了。

Paul the Octopus correctly predicted the winner of all 7 German matches during the 2010 Football World Cup in South Africa, and also correctly predicted Spain as the overall winner. He died shortly after that.

章鱼保罗准确地预测了 2010 年南非足球世界杯赛中德国队全部 7 场比赛的成绩，并且准确预测出西班牙队是冠军。在那之后不久它死了。

Think about It 想一想

Can you imagine the feeling of a piece of an octopus's arm trying to crawl up someone's throat after it had been eaten alive?

你能想象一只章鱼被活吃后，它的一段胳膊爬过人的喉咙的那种感觉吗?

What would your body and head look like without bones or a skull?

如果你的身体或头颅没有骨头，它们会是什么样子?

What other surprises can you imagine from Nature that has, through evolution, developed an animal with eight arms and three hearts?

通过自然进化，一种动物有八条胳膊和三个心脏，你能想象还有更令人惊奇的事情吗?

How fresh do you want to eat food: oven fresh or living fresh?

你喜欢吃多新鲜的食物：刚出炉的还是活鲜的?

Do It Yourself! 自己动手！

Start a search on the internet and look for cultures and traditions around the world where live animals are eaten. The goal is not to accuse or to judge, but to understand that eating live octopus is not an isolated case and that many culinary exploits are based on eating live animals.

在网上发起一个调查，看看世界上哪种文化传统里有活吃动物的习俗。目的不是指责或评判，而是要理解生吃章鱼并不是孤立的案例，很多烹饪上的开拓是建立在生吃动物的基础上的。

学科知识

Academic Knowledge

生物学	在自然界，有规则就有例外；章鱼的眼睛与人的眼睛相似，有角膜、虹膜、晶状体、玻璃体液和视网膜；章鱼是食肉动物，有时候甚至会吃它们的同类；再生能力：海星的肢体切断能够再生一个新海星，但是章鱼的肢体切断只能再生一个新肢体；章鱼有5亿个大脑神经细胞，而人类有1 000亿个大脑神经细胞；章鱼并不从父母那里学习知识，小章鱼出生后它们的父母就死了；章鱼的胳膊是兼具嗅觉和触觉功能的器官，它们可以用胳膊闻气味。
化　学	章鱼的墨汁是由黑色素组成的，与头发和皮肤中的黑色是同样的色素；章鱼的食谱是低热量的海鲜，不但富含钙、钾、磷、硒等元素，而且还含有ω3脂肪酸；章鱼有能折射光线的色素，能使章鱼具有彩虹般的光彩；有些种类的章鱼具有由神经系统控制的发光器，发出光亮的持续时间从不到一秒到几分钟；由于章鱼血液里有一种叫血蓝蛋白的含有铜元素的色素，章鱼的血液是蓝色的。
物　理	章鱼有四对胳膊，需要提供额外的运送血液的压力，章鱼的第三个心脏专职负责向全身提供血液的常规功能；章鱼的幼体漂浮在海面上，捕食它们发现的任何食物，等它长得足够重，就沉入海底继续生长；章鱼不仅能通过色素模拟颜色，而且还能根据环境变化身体形状，把自己隐藏起来，比如可以变得像一块被海藻覆盖的石头。
工程学	通过氧化作用将受污染的水净化成灌溉用水或饮用水。
经济学	鲜鱼和甲壳纲动物市场是价值1 250亿美元的全球性大市场，而藻类市场只有20亿美元，但鱼类市场面临压力，藻类市场仍在增长。
伦理学	什么论据能够支持人类活吃鱼类的欲望？如果我们反对活吃动物，那么我们也应该反对生吃蔬菜和水果吗？
生活方式	韩式活章鱼（一种受欢迎的韩国菜）是把活章鱼切成小块，上菜的时候，章鱼的爪子仍在扭动；日本的活鱼料理是一种由活的章鱼、虾、龙虾切成的生鱼片；在西方文化里，牡蛎是生吃的，比利时人过去常喝杯子里有条活鱼的啤酒。
社会学	活吃动物的传统在亚洲很广泛，但是在欧洲，由于动物权利意识的出现，这种传统不再盛行。
心理学	多数人不仅吃动物肉，而且也关心动物权利，经受着“食肉悖论”的心理挣扎。
系统论	认识到我们被有意识的生命所包围，不只是人类才有知觉。

情感智慧

Emotional Intelligence

牡　蛎

牡蛎没有强化章鱼作为受害者的感觉；相反她证实了其他生物也正遭遇完全相同的境遇。当章鱼坚持强调可怕的未来时，牡蛎一开始嘲笑章鱼有很多胳膊，但紧接着就表达了尊重，最终发展为对其独特性的钦佩。这导致了一场对话。牡蛎对这位朋友越来越感兴趣，对他不断揭示的事实惊讶不已。章鱼以一个关键人物的形象出现，如同科幻电影里的主角一样。牡蛎最终把章鱼视为一个罗宾汉似的英雄，能够与捕猎者一较高下。

章　鱼

章鱼一开始充当了受害者的角色，抱怨所遭受的各种虐待。牡蛎直面章鱼，显露出某种现实主义态度。但听了牡蛎赞美其“腿”的笑话，章鱼并不受用，他平静地纠正牡蛎，但也维护了牡蛎的自尊。当牡蛎奉承章鱼时，章鱼的第一反应是少谈自己的智慧，而是更多地展示各种实用本领。他通过一系列令人惊奇的信息说服牡蛎。他明确清晰地述说他的每一个独有的特征。当牡蛎对章鱼的胳膊再生能力感到好奇时，他不以为然，而是强调了他变化颜色和形体的能力，他把这视为自己最惊人的特征。最终他以轻蔑的口气，完成了从受害者角色到拥有优势地位角色的转换。

艺术

The Arts

画一幅“章鱼风格”的油画。想象你是一只章鱼，你可以随意在白色帆布上喷洒黑色墨水。你将怎样喷墨呢？是用一个细小的高压水针画，还是直接在帆布上泼墨？画几幅水墨画，心理分析师能够从你的画风里了解到你的一些重要的性格特质。当你选择做何种类型的喷墨画家时，你也传递出了你是怎样的人。

思维拓展

Systems: Making the Connections

当学习物理学时，我们发现所有定律几乎都没有例外；学习化学时，我们发现很多情况下，化学反应取决于温度、气压和使用的催化剂；在学习生物学时，我们又发现似乎每件事物都是例外，没有任何事物是完全相同的。不知何故，自然界已经从原始生命、从几个简单的细胞进化为一连串独特的生态系统，有上亿种不同形态的生命形式。生物多样性是神奇的，而其细节的发现，唤起了人们对与我们共享这颗星球的所有生命形态深深的敬意和钦佩。然而，人类似乎经常是无知的，没有觉察到这些难以置信、好像直接来自科幻小说的生命力量。基于这样的背景，活吃动物有违人类不同于其他一切生命形式的基本价值观。对所有生命形态的赞美和尊重意识正在不断提高，同时却有很多人认为动物必须服从于人类。但这就赋予我们虐待动物的特权吗？也许虐待有轻有重，但必须确保所有生物都能有尊严地生存和繁衍。只有每种生物相互尊重和欣赏彼此在这颗星球上的独特性，万物才能生生不息地循环。

动手能力

Capacity to Implement

生命的关键是保持正能量。有太多的理由让人觉得自己是受害者，哀叹自己的不幸。创造新的机会和学习新的知识是很重要的，甚至包括从我们不喜欢的或者让我们感到痛苦的事件中学习。我们有与时俱进的能力，但任何时候，我们都要尊重差异和传统。或许此时，你应该停下来仔细看看你的饮食，看清你在吃什么。你有过“食肉悖论”吗？在你内心，是怎样调节对动物的爱与为获取动物蛋白而吃动物肉的矛盾的？你可以先思考一下这些问题，归纳出正面和反面的理由，然后再与朋友进行辩论，分享你的观点。不管结论是什么，找到一种你良心能够接受，而且吃得舒服的饮食结构。

故事灵感来自

This Fable Is Inspired by

查尔斯·范·德·阿埃让

Charles van der Haegen

查尔斯是10个孩子的父亲，他的妻子为亲生和收养的子女的美好未来奉献了自己的全部身心，这一行为对查尔斯的思想和工作产生了很大影响。查尔斯毕业于商务工程专业，拥有MBA学位。后来，他进入了医疗器械产业。1982年，作为总经理的他被要求关闭仪器仪表公司时，他拒绝了，并把公司转变为比利时首家机器人公司。他从商务经理人转变为投资企业家，带着没有光明前途的公司进入一个竞争激烈的行业。他的职业生涯中不仅有成功，也有失败。正是他顽强的求存精神，让他在困难重重的黑暗中看到光明。查尔斯在商务和社会活动领域贡献着他丰富的经验，特别是作为传递者，在欧洲推行蓝色经济项目。

图书在版编目(CIP)数据

冈特生态童书.第三辑修订版：全36册：汉英对照 /
(比)冈特·鲍利著;(哥伦)凯瑟琳娜·巴赫绘；
何家振等译.—上海：上海远东出版社，2022
书名原文：Gunter's Fables
ISBN 978-7-5476-1850-9

Ⅰ.①冈… Ⅱ.①冈… ②凯… ③何… Ⅲ.①生态环
境-环境保护-儿童读物—汉、英 Ⅳ.①X171.1-49

中国版本图书馆CIP数据核字(2022)第163904号

著作权合同登记号图字09-2022-0637号

策　　划 张　蓉
责任编辑 程云琦
封面设计 魏　来李　廉

冈特生态童书
你真聪明
[比]冈特·鲍利　著
[哥伦]凯瑟琳娜·巴赫　绘
何家振　译